GW00690243

WATER WORKS

SARAH WATTS

DO WE HAVE EQUAL
RIGHTS TO RESOURCES?

KS3 GEOGRAPHY TEACHERS' TOOLKIT
EDITED BY ALAN KINDER AND JOHN WIDDOWSON

Geographical
Association

Acknowledgements

As ever, I am grateful to Mark, for his understanding, constant support and attempts to bring normailty to my life as well as the endless supply of tea and cakes! Special thanks to Guru Fred Martin, for his wise words, belief in me and Friday e-mails. Thanks also to Alan for his patience and encouargement - never again, once was enough!

© Sarah Watts, 2009

ISBN 978 1 84377 214 9
First published 2009
Impression number 10 9 8 7 6 5 4 3 2 1
Year 2012 2011 2010

Published by the Geographical Association, 160 Solly Street, Sheffield S1 4BF.
Website: www.geography.org.uk
E-mail: info@geography.org.uk
The Geographical Association is a registered charity: no 313129

The GA would be happy to hear from other potential authors who have ideas for geography books. You may contact the Publications Officer via the GA at the address above.

Edited by Andrew Shackleton
Designed by Bryan Ledgard
Printed and bound in China through Colorcraft Ltd, Hong Kong

CONTENTS

EDITORS' PREFACE

The *KS3 Geography Teacher's Toolkit* is designed to help teach the new key stage 3 curriculum from 2008. The series draws on the Key Concepts, Key Processes and Curriculum Opportunities in the new Programme of Study and applies these to selected parts of the Range and Content. For teachers, it provides timely guidance on meeting the challenge of creating and teaching the curriculum. Each title in the series illustrates ways of exploring a place, theme or issue of interest to young people and of geographical significance in the twenty-first century. The selection of content is carefully explained, ideas are clearly linked to the new Programme of Study and advice is provided on the use of teaching strategies to engage and challenge all learners in the classroom.

The *Toolkit* can be used in a number of ways. For busy teachers of geography, under pressure from curriculum change throughout the secondary phase, each title in this series provides a complete unit of work: a bank of ready-made lesson plans and accompanying resources. These materials can be used *directly* in the classroom, with minimal preparation. The printed resources in each book may be copied directly, but complete resources for every lesson are contained on the easy-to-navigate CD.

Toolkit materials can also be extended. Each title provides links to websites of interest and to further resources and reading, encouraging teachers and students to 'dig deeper' into their chosen places, themes and issues. Activities within each unit can be extended

into full-scale enquiries, to stretch even the highest attainers.

The *Toolkit* has also been designed to be adapted. Teaching strategies are explained throughout each book, allowing teachers to understand the 'how to' of each lesson activity. It is hoped that teachers using these materials will be encouraged to select ideas, change them to meet the needs of their own learners, and begin to use relevant teaching strategies elsewhere in their curriculum. Each title is therefore a rich source of teacher-to-teacher advice, a 'professional development' resource that can be used to inform the teaching of places, themes and issues of your own choosing.

Lastly, the series provides a template for writing new curriculum materials. Unit summaries, concept maps linked to new Key Concepts, assessment frameworks, glossaries, lesson plans and other materials are included as exemplars of rigorous curriculum planning.

By using, extending, selecting and adapting appropriate 'tools' from the *Toolkit*, teachers will gain confidence in developing their own materials and creating a high-quality curriculum suited to the needs and interests of their learners. We hope that the series will help teachers fully exploit the rich potential of the new KS3 Programme of Study.

Alan Kinder and John Widdowson, 2008.

I. WATER WORKS: DO WE HAVE EQUAL RIGHTS TO RESOURCES?

Why teach about water rights, conflicts and management?

Water is a vital resource which we all need and use

Teaching about water supply issues should help learners to understand how important water is. The use of water impacts on all our lives, from our health to the production of the food we eat. Secondary school students in the UK simply turn a tap in order to access a constant supply of fresh, safe (and seemingly free!) water (Figure 1). They have little comprehension of what life would be like without this supply. To really appreciate what a valuable resource water is, learners need to consider what we mean by water rights and water scarcity.

The availability of clean water is an issue of growing global concern

While the world's population tripled in the twentieth century, the use of water resources grew six-fold. However, the amount of water on the earth's surface is fixed. Within the next 50 years, world population is projected to increase by another 40-50%. This population growth, coupled with industrialisation and urbanisation, will increase demand and make water an increasingly important issue for today's learners.

In some places water is being used unsustainably and without responsibility

Water is essential to life and as such is a basic human need, but we don't all have equal access to this essential resource. In countries like the UK, clean and reliable water is now regarded as a human right (Watkins, 2006). Learners need to appreciate their right to water and not abuse it. They need to recognise the responsibility (of individuals, households, organisations and governments) to use water sensibly and sustainably, and take action to use water responsibly themselves.

Water can force co-operation in some areas and cause conflict in others

Water is a resource that crosses borders: 145 nations have territory within a 'transboundary' river basin

Figure 1: What is easily taken for granted here means life or death for people elsewhere in the world.
Photo © Kantor Zsolt/Morguefile.

(covering more than one country), and 21 lie entirely within one. While agreements over the use of water, such as the Colorado Compact in the USA, are not new, water stress underlies many conflicts around the world: Israel, Jordan and Syria have battled since the 1950s for more effective control of their sparse water resources. Even where countries are not involved in military conflict, tensions about how to share scarce resources are present wherever rivers are shared (Clarke and King, 2004). There seems every possibility of water being a source of serious future conflict between countries.

The Middle East is a region of vital importance – yet is rarely studied in key stage 3

With 5% of the world's population surviving on 1% of its water, there is strong competition for water in the Middle East. The region is often characterised as war-torn, with learners' perceptions being formed predominantly by a media focus on religious conflict. This unit helps learners to see the region from a new perspective, to begin to understand some of its complexity and to recognise some of the ways in which daily life might be very similar to, and different from, life in the UK.

The use and abuse of water raises questions of rights and responsibilities, power and poverty, consumption and sustainability, even life and death. It makes planet Earth blue, green and populated. It is geography at its most basic – a drop of 'real geography'!

How to teach about water rights, conflicts and management

Learn through talk

People hold different viewpoints and one of the best ways these can be considered and understood is through discussion. Geography is about people, their opinions and views. When learners voice their own views and listen to others, they begin to understand that sometimes there is no answer that is equally 'right' for everyone. They begin to recognise the similarity between their own classroom discussions and the negotiations (and disagreements) taking place between countries striving to agree on future water supply and management. As with any real-world issue, there is more than one way of looking at water rights, conflicts and management. Through talk learners formulate and clarify their ideas about water use and conservation, and develop their thinking in relation to complex questions such as 'Do we have a right to water?'.

Think critically

At its best, geography in key stage 3 promotes an active and critical approach to enquiring about the world. In geography, students need to be given opportunities to solve problems and make decisions in order to develop analytical skills. In this unit, students are encouraged to collaborate, think, discuss and reflect together in an inclusive fashion. Critical and creative thinking about geographical issues is promoted through strategies such as mystery activities. Each lesson in the unit is posed in terms of a question to encourage thoughtful enquiry in relation to lesson objectives.

Examine geographical issues in the news

The relative prominence of the Middle East on our television screens is certainly not matched by the number of geography lessons given over to it. Are the images of the region shown in the UK media accurate? This unit acknowledges the unique challenges and conflicts over the use of water in the region. It considers predictions about future 'water wars' and emphasises the need to enquire and find out more. It challenges learners to form their own views, to give reasons and justify their viewpoints.

Build from personal perspectives

Personalised learning is about making learning experiences relevant and applicable. From this approach, students begin to think critically, constructively and creatively, using themselves as the starting point. This provides a platform for examining and understanding the unfamiliar. In order to make informed decisions about water use, it is important to engage students by demonstrating their hidden reliance on water. By asking thought-provoking questions such as 'Do you know that it takes over 4000 litres of water to produce a cotton T-shirt?', learners reflect on their own uses of water and begin to generate their own questions, such as 'How does my use of water compare with that of people around the world?'.

Summary

This is a unit about water resource use, conflict and management. Within this unit learners are encouraged to understand the consequences of their own use of water and appreciate how important it is to manage this essential resource. They contrast their own experiences with those living in the Middle East where access to a constant and clean water supply provides a challenge. They are given opportunities to decide for themselves how they would manage water in the future and make informed decisions on their personal water consumption.

The investigation allows learners to develop concern and understanding for the environment and to understand how their actions can influence their future. There are significant links with citizenship (e.g. human rights and rights to water).

Prior learning

In key stage 2 students learn about water and its effects on landscapes and people, including the physical features of rivers or coasts. They may have explored water supply around the world, learned about the importance of clean water, investigated how it is supplied and contrasted water use at home with its use in less economically developed countries. Learners in key stage 2 also study conflicts over land use, for example through studying national parks or coastlines.

Future learning

This unit develops an understanding of the complexity of resource use. Skills gained from this unit could be developed when examining the management of other resources, e.g. energy resources.

Learners could deepen their study of hydrology, and examine the hydrosphere and the systems operating within it. Flooding and river basin management could be studied, as could drought and desertification. The theme of water supply could be further developed by looking at the agencies involved in water supply across the globe, or at the impacts of different models of water aid.

Key learning outcomes

Most students will be able to:
- describe the importance of water to people's quality of life
- explain patterns in the supply of, and demand for, water
- express their own and others' views on water resource conflicts.

Some students will not have made so much progress and will be able to:
- recognise the importance of water in their own lives
- show awareness of uneven patterns of water supply
- express their own views on water conflict and management.

Some students will have progressed further and will be able to:
- explain why water rights and responsibilities are not evenly applied across the world
- appreciate different approaches to solving water resource conflicts.

The geography behind Water Works

'We live within a system of finite resources. There isn't much we can do to alter the actual quantity of water on the earth. In a finite liquid cycle, the sun's energy sucks up water from the earth and

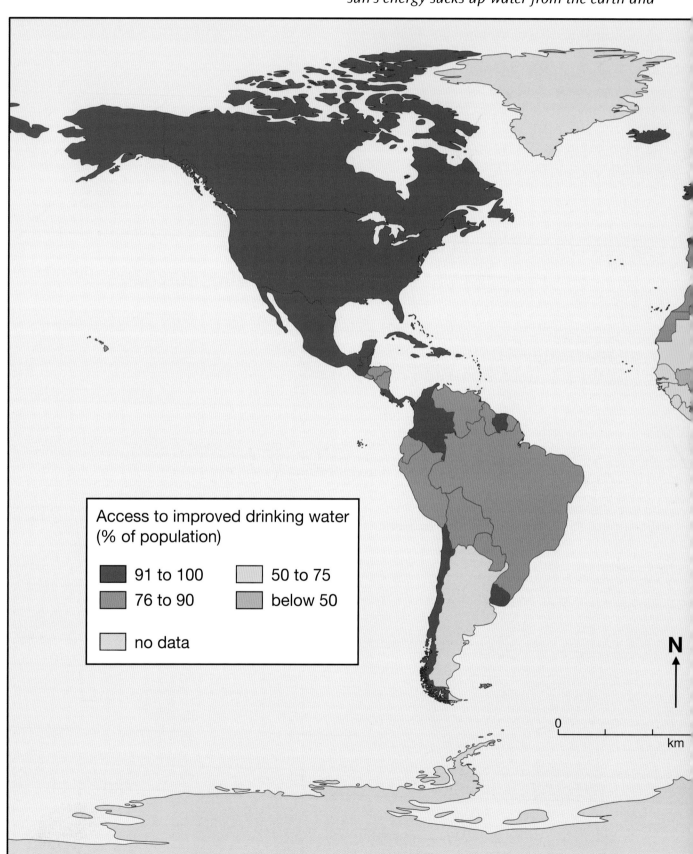

Access to improved drinking water (% of population)

- 91 to 100
- 76 to 90
- no data
- 50 to 75
- below 50

N

0   km

Figure 2: Population without access to clean water. Source: Longman Student Atlas, 2005.

sends it back again as rain, sleet, or snow. But although this supply of water is largely fixed in amount, a great deal can be done to alter its location and quality' (Raines Ward, 2002).

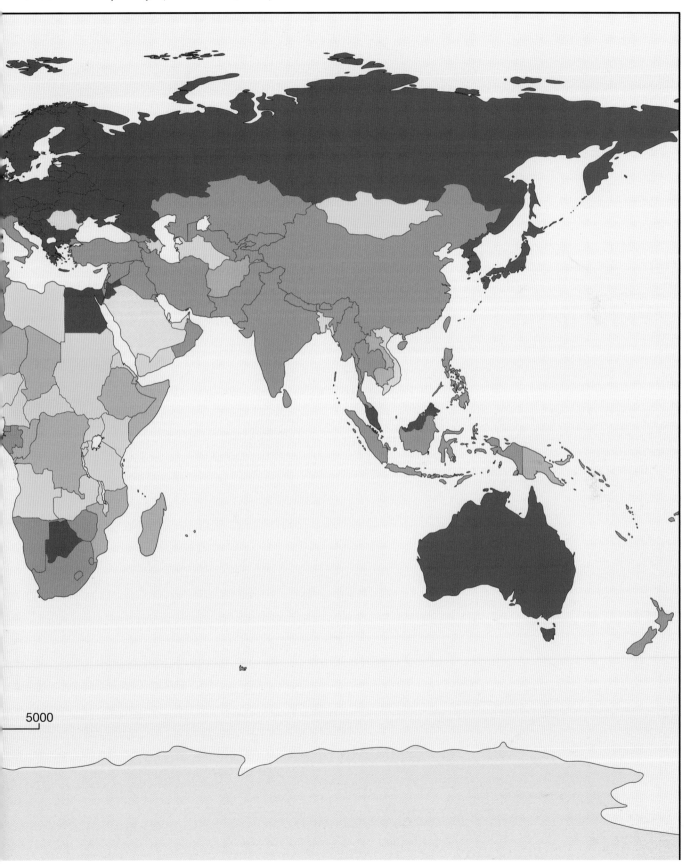

5000

Water Works: Concept map

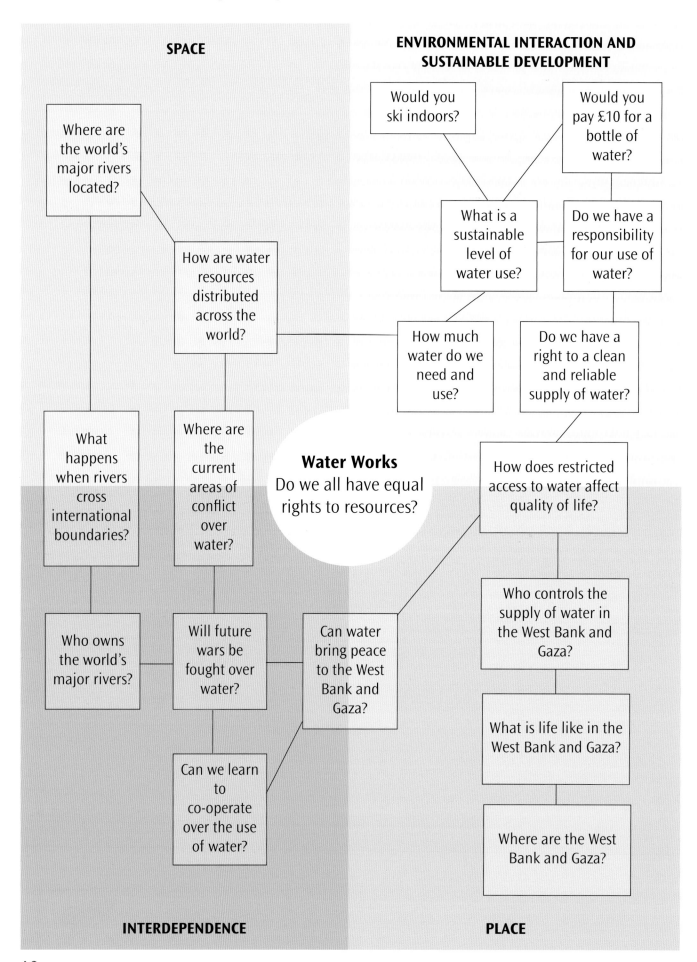

SPACE

ENVIRONMENTAL INTERACTION AND SUSTAINABLE DEVELOPMENT

Where are the world's major rivers located?

How are water resources distributed across the world?

Would you ski indoors?

Would you pay £10 for a bottle of water?

What is a sustainable level of water use?

Do we have a responsibility for our use of water?

How much water do we need and use?

Do we have a right to a clean and reliable supply of water?

What happens when rivers cross international boundaries?

Where are the current areas of conflict over water?

Water Works
Do we all have equal rights to resources?

How does restricted access to water affect quality of life?

Who owns the world's major rivers?

Will future wars be fought over water?

Can water bring peace to the West Bank and Gaza?

Who controls the supply of water in the West Bank and Gaza?

Can we learn to co-operate over the use of water?

What is life like in the West Bank and Gaza?

Where are the West Bank and Gaza?

INTERDEPENDENCE

PLACE

Links to the National Curriculum Programme of Study

Key concepts

Place
This unit:
- promotes understanding of the physical and human characteristics of the Middle East as a region in which the demand for water is increasingly outstripping supply
- develops geographical imaginations of the Middle East beyond the perceptions commonly formed from images in the news and other media.

Space
This unit:
- explores the interactions between places and the networks created by the transfer of water from areas of surplus to areas of deficit
- locates places of water surplus, scarcity, co-operation and conflict.

Scale
This unit:
- investigates the distribution and consumption of water resources at a range of scales
- makes links between scales to develop an understanding of the personal, national and international dimensions relating to water resources.

Interdependence
This unit:
- explores the nature and causes of conflict and co-operation between places, arising from pressure on water resources
- explains the significance of interdependence in water supply, distribution and consumption.

Environmental interaction
This unit:
- shows how the physical and human dimensions of water supply and consumption are interrelated and together influence environmental change
- describes sustainable and unsustainable uses of water resources.

Key processes

Geographical enquiry
This unit:
- is structured around key questions and problems that students investigate and solve
- requires students to make decisions about water use, management and supply.

Graphicacy and visual literacy
This unit:
- uses varied resources, climate data, atlases and maps
- encourages students to interpret images of places and features and infer meaning from them.

Geographical communication
This unit:
- provides opportunities to communicate knowledge and understanding of water management, supply and sustainable use
- requires students to learn and use geographical vocabulary about water
- develops analytical and creative thinking skills about contentious issues.

Range and content
This unit:
- makes connections at a range of scales, from personal to local, regional and international
- shows the interactions between people, water and their environment
- investigates the issue of water supply, management and sustainable use.

Curriculum opportunities
This unit:
- explores real and relevant contemporary issues of water management
- makes curriculum links between geography and citizenship
- requires students to reflect on their own actions and anticipate the consequences of these
- examines views held about locations
- encourages learners to consider different views on water supply, demand, management and sustainable use
- uses a range of enquiry approaches
- uses varied resources, including maps and visual media
- examines geographical issues in the news.

2: WATER WORKS:
Medium-term plan

Lesson	Key questions	Learning objectives	Teaching and learning	Resources	Assessment opportunities
1	How much water do we use or need? What do we use water for? How does water affect our lives?	To recognise ways in which we consume water To understand the importance of water to quality of life To be able to estimate personal water consumption	Focus on how much water we use and need Students identify the connections between a series of images related to the theme of water Students examine their own water use, estimate their water consumption and consider patterns of consumption around the world. Students consider virtual water consumption (water used to produce consumer products) Students identify themes and questions relevant to the remainder of the unit	Figures 3-15 from CD Activity Sheet 1 Activity Sheet 2 Video clip of 'A World Without Water' from http://video.google.com/videoplay?docid=3930199780455728313	Students estimate their daily use of water and the water needed for consumer products they use Students identify areas of high and low water consumption across the world Students predict the impact of very low water availability
2	Where in the world is all the water? What is the pattern of water resources across the world? Why do some places enjoy a water surplus and others suffer water deficit?	To know a range of sources of water To know that water resources are unevenly spread across the globe To understand factors that create a water surplus or deficit	Focus on water availability across the world Students match countries with their capital cities and some water features Students identify a range of sources of water and ask, 'Is there enough water for us all?'. They plan how to answer this question and investigate the water balance in one country as part of their enquiry Students use the 'Where am I?' questioning technique to identify countries referred to in the lesson	Activity Sheet 3 Information Sheet 1 Activity Sheet 4 Activity Sheet 5 Figure 16 from CD Atlas Sticky notes	Students identify different sources of water and decide which are accessible to people Students use evidence to predict whether a country has a water surplus or deficit Students identify and describe patterns of water supply and demand
3	Do we all have a right to water? Do we have a right to a clean and reliable supply of water? Do we have responsibilities regarding our use of water?	To understand the link between water and basic human rights To recognise that with water rights come responsibilities	Focus on water rights (rights of access, quality and quality) Students play a game of dominoes – distinguishing between wants and needs Students explain links between universal human rights and the availability of water. They identify irresponsible uses of water (including their own)	Activity Sheet 6 Video clip 'Water: a drop for life' (www.un.org/waterforlife decade) Information Sheet 2 Information Sheet 3	Students identify a range of human wants and needs and distinguish the two Students suggest how lack of water can impact on other human rights Students identify irresponsible uses

Lesson	Key questions	Learning objectives	Teaching and learning	Resources	Assessment opportunities
3 cont.			Students discuss some 'philosophical' questions such as, 'Do I have the right to unlimited water use?'	Activity Sheet 7	of water and ways of conserving water
4	Who owns the world's water? What happens when rivers and sources of water are shared between countries? What impacts can building dams have?	To know where countries co-operate or conflict over major rivers and water sources To understand different points of view about who owns rivers and water sources	Focus on water resources in the Middle East Students identify examples of conflict and co-operation with which they are familiar Students examine the Euphrates and Tigris river basins and the disagreements between Syria, Turkey and Iraq over water. They consider different viewpoints and decide whether the Ilisu dam project should go ahead Students consider who 'owns' the water in rivers, and what responsibilities this ownership brings	Information Sheet 4 Information Sheet 5 Information Sheet 6 Activity Sheet 8	Students express a viewpoint on a new dam scheme Students decide whether a large new dam scheme should go ahead
5	What is life like without water 'on tap'? How can restricted access to water impact on quality of life?	To understand how restricted access to water supply can impact on quality of life To interpret, analyse and sort evidence and draw conclusions	Focus on the use of water resources in the West Bank region Students take part in a 'Where am I?' quiz They work in pairs to solve a mystery: 'Why does Ahmed need a new pair of shoes?'. They analyse information and make connections between pieces of evidence to reconstruct the story of his day spent collecting water Students debrief on how they solved the mystery. They take part in a 'human continuum' to express their opinion on a range of solutions to water poverty in the West Bank	Activity Sheet 9 Activity Sheet 10 Activity Sheet 11 Information Sheet 7	Students interpret evidence to explain the effects of water inequality on one family Students express opinions about a dispute over water resources
6	Is water used fairly in Israel and the West Bank? How do Israelis and Palestinians	To know the location of key features in the Jordan River basin To understand that water supplies in the region are	Focus on the supply and conflict over water within the Jordan River basin Students locate the Jordan River in an atlas and identify key geographical features of the region. They add information to their own maps and overlay these to show the location of	Activity Sheet 12 Activity Sheet 13 Information Sheet 8	Students annotate a map of the region with geographic information Students explain how Israelis/Palestinians

Lesson	Key questions	Learning objectives	Teaching and learning	Resources	Assessment opportunities
6 cont.	feel about the unequal supply of water? Should people always have equal access to water?	unequal To be able to argue for/against a point of view	underground aquifers Students create a 'fortune line' to establish a basic chronology of water use in the area. They consider how both Israeli and Palestinian people might respond to a range of events and information about water resources Students suggest ways in which conflicts over water resources could be resolved	Information Sheet 9 Figure 17 from CD Atlases	might feel about water supply issues
7	Is skiing a good use of water? How can we assess sustainability? Why do people hold very different opinions about the use of water resources?	To know the meaning of the word 'sustainable' To understand that sustainability includes economic, social and environmental elements To be able to assess the sustainability of water use from different points of view	Focus on the development of indoor skiing centres Students create a definition of 'sustainable development' They use 'layered decision making' to rank the location factors for a ski resort. They guess the location of a resort from information about it, and consider its economic, social and environmental impacts from different points of view Students reflect on why indoor ski resorts are being built, their good and bad features and whether they would choose to ski indoors	Information Sheet 10 Information Sheet 11 Information Sheet 12 Information Sheet 13 Activity Sheet 14 Figures 18-27 Sticky notes	Students complete a diagram to compare points of view on a ski resort Students assess the sustainability of indoor skiing and explain why resorts are being built
8	What is the cost of a bottle of water? How much would you pay for a bottle of water? What are the impacts of drinking bottled water?	To analyse the reasons people drink increasing quantities of bottled water To understand the impacts of drinking bottled water	Focus on the costs of increased bottled water consumption and production Students analyse adverts for bottled water and summarise the messages conveyed They learn about the growth of bottled water and identify 'rogue data' shown to them Students investigate the 'cost' of a bottle of water by investigating popular and luxury brands, as well as the environmental impacts of bottled water production, transportation and disposal. Students design their own message to convey their response to the question: 'Are we paying too much for bottled water?' One student takes part in a taste test to distinguish chilled tap water from bottled water	Bottles of water and notes on advertising brought in by students Cups or glasses One glass of chilled tap water Video clip 'Bottled water: is it worth it?' (http://news.bbc.co.uk/player/nol/newsid_7230000/newsid_7231200/7231261.stm?bw=bb&mp=wm&news=1&ms3=6&ms_javascript=true&bbcws=2)	Students identify bias and the use of persuasion in a range of images and other sources Students assess the social, environmental and economic costs and benefits associated with bottled water
6 cont.					

Lesson	Key questions	Learning objectives	Teaching and learning	Resources	Assessment opportunities
				Blindfold or scarf Information Sheet 14 Information Sheet 15 Information Sheet 16	
9	Will our water run out? Who supplies my water? What do I think should happen to water use in future?	To know how water is supplied and consumed in the UK To understand the need to balance water supply and demand To imagine a future with high or low levels of water use	Focus on the balance of supply and demand for water in the UK in the future Students read statements about water rights, responsibilities and management and state how far they agree with these Students examine water company mission statements and identify issues affecting water supply and demand. They write about one future scenario for water use in their area Students write their own 'message in a bottle' suggesting their preferred future for water. They read one another's messages	Activity Sheet 15 Information Sheet 17 Information Sheet 18 Activity Sheet 16	Students rank statements about water use and write their own statement for the future Students write a letter describing a possible future for water resources

LESSON 1:
How much water do we use or need?

Key questions
- What do we use water for?
- How does water affect our lives?

Key words
- resource
- virtual water
- water footprint
- consumption
- quality of life

Resources
- Figures 3-15 from CD
- Activity Sheet 1
- Activity Sheet 2
- Video clip of 'A World Without Water' from http://video.google.com/videoplay?docid=3930199780455728313

Learning objectives
- To recognise ways in which we consume water
- To understand the importance of water to quality of life
- To be able to estimate personal water consumption

Assessment opportunities
- Students estimate their daily use of water and the water needed for consumer products they use
- Students identify areas of high and low water consumption across the world
- Students predict the impact of very low water availability

Lesson structure

Starter
Before revealing the unit title, lesson title or objectives, tell students they are going to see a selection of images – they need to decide what connects them! Show Figures 3-7 from the CD and ask students to offer their views. Show Figures 8 to 11, then ask again. After showing Figures 12-15, ask for a third set of answers.

> 'Confident uncertainty' means being comfortable offering ideas and hypotheses. Not all students will be confident at making suggestions to the whole class. Use pairs, and lots of encouragement, to promote student confidence. These images are connected by the theme of water. There may be many other connections between them that students can identify.

Main teaching phase
In groups, give students a copy of Activity Sheet 1 and ask them to identify how they use water and to agree an estimate of how much water they use per day (in litres). Record the estimated total from each group and share some of the common direct uses of water (e.g. drinking, washing etc.).

When the students have completed questions 1-5, ask what they consider to be the 'minimum' water need without which their quality of life would be seriously affected? What would happen if they were denied access to this amount of water? Read out the following statistics:

- The average person in the UK uses 150 litres of water a day
- Direct water consumption includes 2 litres for cleaning teeth, 25 litres to shower, 90 litres to run a bath, 9 litres per toilet flush and 2 litres to drink
- The UN recommends a minimum of 50 litres a day per person: 5 for drinking purposes, 20 for sanitation, 15 for bathing and 10 for cooking.

Ask students where rates of water consumption are likely to be highest and lowest around the world. Are there likely to be any countries where consumption is lower than the recommended minimum? How might people's lives be affected there?

Explain the difference between *direct* and *virtual* water consumption. Ask students to complete the table estimating how much water it takes to produce the different items. Compare the students' estimates with the actual totals below then ask the groups to re-estimate their total water use, including their *virtual* water consumption.

> A water footprint is the total amount of water an individual consumes, both directly and indirectly. It includes the volume of water that is used to produce the goods and services consumed by individuals – the virtual or embedded water.

Answers to Activity Sheet 1

Country	Average water use per person per day (litres)
UK	150
France	280
USA	575
Brazil	180
Kenya	60
Spain	325
Mozambique	5

Item	Number of litres of water it takes to produce
One kilo of beef	24,000
One hamburger	200,000
One ice-cream	1500
One glass of orange juice	1700
One apple	70
One pint of beer	170
One cup of tea	35
One pair of jeans	11,000
One car	400,000

Domestic water use is only one element of our total daily water consumption. Water is used in the production of goods including food, clothes and cars. This is called virtual, or embedded, water use. Even if students are efficient in their personal water consumption, when they include the water used in growing crops and rearing livestock for their food, and production of their clothes, consumables and electronics, their water use increases dramatically.

Plenary

Show students the video clip 'A world without water'. What issues about water does it raise?

Ask students what they feel a unit on water supply issues should include. What are the likely themes and questions they will need to tackle in the rest of the unit? They record their responses on the KWL grid on Activity Sheet 2.

> A 'KWL' grid is a useful way of engaging students in the enquiry process at the beginning of unit. It encourages them to share things they think they already know (K), things they want to find out (W) and things they have learned (L) by the end of the unit. In this lesson, it could be used to summarise things they have learned in the main phase.

Photo © Ronnie Bergeron/Morguefile

LESSON 2:

Where in the world is all the water?

Key questions
- What is the pattern of water resources across the world?
- Why do some places enjoy a water surplus and others suffer water deficit?

Key words
- hydrosphere
- water cycle
- water balance
- surplus
- scarcity
- groundwater
- soil moisture
- ice sheet
- glacier

Resources
- Activity Sheet 3
- Information Sheet 1
- Activity Sheet 4
- Activity Sheet 5
- Figure 16 from CD
- Atlas
- Sticky notes

Learning objectives
- To know a range of sources of water
- To know that water resources are unevenly spread across the globe
- To understand factors that create a water surplus or deficit

Assessment opportunities
- Students identify different sources of water and decide which are accessible to people
- Students use evidence to predict whether a country has a water surplus or deficit
- Students identify and describe patterns of water supply and demand

Lesson structure

Starter
Introduce the 'Water, Capitals and Countries' heads and tails activity to be completed as a whole-class activity. Give each student two random cards from Activity Sheet 3 then start the activity by calling out 'England' to the class. The student who thinks he or she has the capital of England on their card should call out the answer ('London'), then call out what is on their second card. The student that thinks he or she has the matching card calls out their answer, then calls out what is on their second card, and so on. Care should be taken as some of the heads and tails relate to water – not capitals!

> This is a 'quick fire' whole-class starter, designed to engage student interest and get everyone involved. It also allows the teacher to gauge the students' locational knowledge in relation to the countries used in the main phase of the lesson. Students should retain their country cards for later in the lesson. For your information, the pairs are shown below.

London	
Australia	Canberra
Israel	Jerusalem
France	Paris
Japan	Tokyo
Canada	Lake Superior
Italy	Rome
Kenya	Lake Victoria
Morocco	Rabat
Peru	Lima
Russia	Moscow
Spain	Costa del Sol
Norway	Oslo
Hungary	Budapest
Thailand	Bangkok
China	Beijing
Mexico	Mexico City
Gambia	Banjul
India	New Delhi
Turkey	Ankara
Egypt	River Nile
United Arab Emirates	Dubai
Greece	Athens

Dublin	River Liffey
Ethiopia	✓Addis Ababa
Brazil	✗ River Amazon
✗River Mississippi	USA
Zambia	Lusaka

Main teaching phase

Explain that the hydrosphere is 'that part of the earth that contains water'. Brainstorm with the class sources or stores of water across the world (e.g. rivers, oceans, rain, ice caps). Distinguish between the three states of water – liquid, solid (ice) and gas (vapour) – and between fresh and salt water. Give students Activity Sheet 4 and ask them to identify which sources of fresh water are easily available to people. Show the map of world water resources (Figure 16) either on a whiteboard or as a handout. What does this teach us about the supply of water?

> Note that only 2.5% of the world's water is fresh, and that more than two thirds of this is unavailable for human use. The global supply of fresh water is therefore less than it first seems, hence the riddle 'Water, water everywhere, and not a drop to drink!'. High attainers might also begin to wonder whether water is always available where and when it is needed.

Ask, 'Is there enough water for us all?'. What evidence is needed in order to answer this question? Give the class an example of one piece of evidence to get them started (e.g. rainfall).

> The ability to ask meaningful geographical questions and identify sources of evidence is an essential element of student enquiry. Organise the activity as a 'snowball'; allowing pairs of students time to think before they form groups of four, then eight, in order to refine their ideas.

Elicit from the class concepts such as rainfall, evaporation, water use, population and river flow. If necessary, refer to the 'Water balance' diagram on Information Sheet 1.

How does water balance vary from country to country? Hand out cards from Activity Sheet 5 to pairs of pupils and demonstrate how to locate a place in the atlas and find major water sources such as lakes or rivers. Model how to use the climate data and the atlas evidence to predict if it:

- has higher or lower water consumption than the UK (link back to lesson 1)
- depends on water originating outside its territory
- has a water surplus or water deficit.

Having located their country and its major water sources, ask students to predict the water balance for their country. Allow them to look at the map of freshwater availability (Figure 16) to check the accuracy of their prediction.

> Demonstration is a key teaching skill. It shows students how to execute a technique and provides them with advice on best practice – in this case, the use of an atlas index and key. Modelling makes the thinking process explicit to students. The teacher 'thinks aloud' as they tackle a technique or problem, in this case interpreting evidence.

Plenary

Ask a student to write the name of one of the countries used in the lesson onto a sticky note. Without reading it, place the note on your forehead and model the 'Where am I?' questioning technique. The aim is to discover what country is written on the label by asking questions that can be answered only 'yes' or 'no'. For example, 'Am I an area under water stress?' 'Am I in Africa?' 'Do I have a major river running through me?'.

Homework

Give students a copy of Figure 16 and ask them to pick out and name specific areas of surplus and scarcity, suggesting reasons for this.

Photo © Bryan Ledgard

LESSON 3:

Do we all have a right to water?

Key question
- Do we have a right to a clean and reliable supply of water?
- Do we have responsibilities regarding our use of water?

Key words
- need
- want
- human right
- responsibility
- United Nations (UN)
- citizen
- conserve

Resources
- Activity Sheet 6
- Video clip 'Water: a drop for life' (www.un.org/waterforlifedecade)
- Information Sheet 2
- Information Sheet 3
- Activity Sheet 7

Learning objectives
- To understand the link between water and basic human rights
- To recognise that with water rights come responsibilities

Assessment opportunities
- Students identify a range of human wants and needs and distinguish the two
- Students suggest how lack of water can impact on other human rights
- Students identify irresponsible uses of water and ways of conserving water

Lesson structure

Starter
Ask students to play the 'Wants and Needs' matching game in pairs. Provide each pair with a copy of Activity Sheet 6 and ask them to consider if each item is a want or a need then to write them under the appropriate category. When the game is completed, elicit a list of 'needs' from the class and use these to help distinguish a 'need' from a 'want'.

Main teaching phase
Introduce the idea of human rights. Explain that these are things that are essential for us to live – basic standards without which people cannot live in dignity. They are universal. Explain that water was not an original UN human right but without it many of our other rights cannot be met (Information Sheet 2 provides background information). Ask students to identify some of the ways in which lack of access to fresh water and sanitation might impact on human rights. As a stimulus show Information Sheet 3 on the whiteboard or as a handout.

> Water was not an explicit right in the original UN human rights conventions (1948). However, Article 24 of the 1989 United Nations Convention on the Rights of the Child states that 'every child should have the right to clean drinking water'.

Introduce the idea that with rights come responsibilities. Play the 'Water: A drop of life' clip from the UNESCO website as a stimulus. Ask students to come up with responsible and irresponsible uses of water – thinking about their own consumption as well as on a global scale. Ask students to think about their typical day, and ways in which they could conserve water.

> Focus on the quality of discussion for this exercise - avoid drawing up tables and asking students to record their thoughts in writing.

Plenary

Give students some 'philosophical' questions about water to discuss:

- Do I have the right to unlimited water use?
- Do I have the right to free water?
- Do I have the right to have my dirty water removed for me?
- Do I forfeit my right to water if I abuse it, e.g. overuse it or pollute it?
- Is it right for people in water-rich countries to waste water when others have so little?

The 'Philosophy for Children' approach encourages the posing of open and challenging questions for students of all ages and levels of attainment. Use the plenary discussion cards from Activity Sheet 7 and allow students to take part in an 'online forum' style discussion. The SAPERE website offers further advice for organising this type of discussion.

Homework

Ask students to create an illustrated slogan for water rights and responsibilities. Give the example 'Water – you know it's right'. Some students may be able to make up a simile, e.g. 'Water *is like* diamonds: clear, valuable and precious' or even a metaphor, e.g. 'Water *is* life'.

LESSON 4:
Who owns the world's water?

Key question
- What happens when rivers and sources of water are shared between countries?
- What impacts can building dams have?

Key words
- conflict
- co-operation
- drainage basin
- dam
- watershed
- water dependency

Resources
- Information Sheet 4
- Information Sheet 5
- Information Sheet 6
- Activity Sheet 8

Learning objectives
- To know where countries co-operate or conflict over major rivers and water sources
- To understand different points of view about who owns rivers and water sources

Assessment opportunities
- Students express a viewpoint on a new dam scheme
- Students decide whether a large new dam scheme should go ahead

Lesson structure

Starter
Introduce and explain the terms 'co-operation' and 'conflict'. Ask students to brainstorm different types of conflict and co-operation from their own experience in life, e.g. with other students, with parents, playing sports. Ask them to add in other examples they know about from the media, e.g. large-scale conflicts between countries. Use Information Sheet 4 to point out areas of water conflict around the globe.

> According to UNESCO, the past half-century has witnessed more than 500 conflicts over water, seven of which have involved violence. As many as 260 river basins in the world transcend national borders and 145 nations have territory in a river basin that crosses borders. Water is predicted by some to be a key cause of future conflicts around the globe.

Main teaching phase
Arrange students in groups of four or five and give each group a copy of page 1 of Information Sheet 5 showing the Euphrates and Tigris river systems region. Allow students some time to examine the map and explain why conflict over water supply is likely. Page 2 of the Information Sheet has some supporting information for teachers.

> Support groups by suggesting factors that might cause potential conflict over access to water, e.g. major rivers crossing national boundaries, arid zones and areas with limited resources. Encourage students to find appropriate map evidence for this.

Explain that a dam controls water flow and thus has impacts at its site and further downstream. Discuss the positive and negative impacts of a dam on the landscape, water supply, power supply and river flow. Provide students with Information Sheet 6. Arrange students into three groups representing each country affected, giving each group their relevant card from Activity Sheet 8, and ask each group to construct a short speech about the scheme, to be given at a meeting of government ministers from Turkey, Iraq and Syria. They should decide which points from the dam decision cards are most powerful, and pose some awkward questions to their neighbours! Re-arrange students into groups of three, with one representative for each country, and ask them to role-play the character on their card.

Ask each group to make a decision about the future of the Ilisu dam scheme. Ask them to consider:
- the impacts on each country
- other major water supplies in the area
- ways of altering the scheme to resolve any conflict.

Plenary

Explore the following questions with the class:

Who 'owns' the water in a major world river? If people have a *right* to the water that lies within their own territory, what *responsibilities* do they have towards others using it further downstream?

> Build on the use of talk earlier in the unit. Encourage students to speculate, give reasons for their views and seek them from others. Ensure that reasoning is made explicit in the talk. Highlight the need to be critical but constructive of others' views. Allow students to take the lead and avoid supplying answers and conclusions yourself.

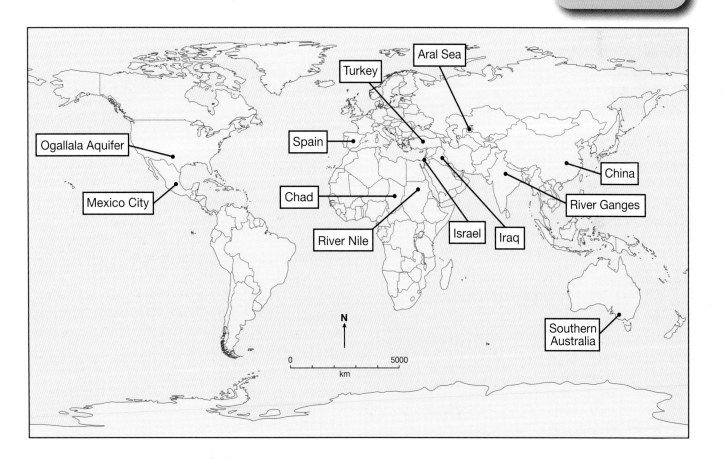

LESSON 5:

What is life like without water 'on tap'?

Key questions
- How can restricted access to water impact on quality of life?

Key words
- spring
- peace accord

Resources
- Activity Sheet 9
- Activity Sheet 10
- Activity Sheet 11
- Information Sheet 7

Learning objectives
- To understand how restricted access to water supply can impact on quality of life
- To interpret, analyse and sort evidence and draw conclusions

Assessment opportunities
- Students interpret evidence to explain the effects of water inequality on one family
- Students express opinions about a dispute over water resources

Lesson structure

Starter
Read the 'Where am I?' quiz questions one at a time from Activity Sheet 9 and ask students to guess the country. Allow several guesses after each clue and encourage students to explain their answers. The answer is Israel.

Emphasise that this lesson will require the use of evidence and good reasoning skills.

> This strategy allows the teacher to gauge the hypothesising and reasoning skills of students. These skills are vital in the main phase of the lesson. The information on Israel also provides a contrast with the life of a Palestinian highlighted in the lesson. A visual alternative to this starter could be created using photographs of the country.

Main teaching phase
Put the students into small groups and share the mystery title 'Why does Ahmed need a new pair of shoes?' with the class. Explain that mystery activities require skills of interpreting and sorting evidence, hypothesising, problem-solving, reasoning and explaining, and that collaboration is essential. Students will need to speak and listen carefully and resolve conflicts within their group.

Distribute a set of cards from Activity Sheet 10 to each group and tell them to begin the activity by reading the statements. Teachers' notes are provided on Information Sheet 7.

> Some text might be difficult to access for some students. Use grouping as means of differentiation and ask each student to take a turn to read, supported by the others, and to work together to understand the information.

Give time and support for students to sort through pieces of evidence and find links between them. They should decide for themselves how to categorise the statements, generate their own 'hypotheses' and test these against the evidence.

Elicit from students their solutions to the mystery. Encourage students to give extended explanations, and invite others to challenge answers that lack evidence or miss out key pieces in the puzzle. Why doesn't he simply use the well, or buy water? What impacts would reduced access to water have on Ahmed and his family?

Next, focus on how they approached the task and, in particular, how they categorised and linked the statements. Ask questions such as 'How did you sort the cards?' and 'How did you agree a conclusion?'.

> Metacognition encourages transference of learning to other contexts. It involves discussing and making explicit the thinking strategies and processes used by students. Highlight the sifting and linking of evidence, the construction and testing of hypotheses. Ask 'What other situation/subject could you use these skills in?'.

Plenary

Use a human continuum strategy.

Pose the following statements, and ask students to arrange themselves on the scale in response to the following statements:

- Israelis in Iraqburin have the right to extract as much water from the spring in their village as they wish.
- Palestinians in Madama should 'get even' by contaminating the Israeli water supply in Iraqburin.
- The United Nations should decide how much water Israelis and Palestinians get – the two sides will never agree.

A human continuum is a scale of opinion – from 1 (strongly disagree) to 5 (strongly agree), with 3 being 'unsure'. Place pages 1-5 from Activity Sheet 11 along a wall of the classroom. Ask students to place themselves by the one that reflects their view in response to statements they will be given.

Ask students to justify their place on the scale and hence their viewpoint.

Homework

Ask students to write a diary or a blog describing how water affects daily life in a Palestinian village, or to record their own use of water for a week.

Photo © Kantor Zsolt/Morguefile.

LESSON 6:
Is water used fairly in Israel and the West Bank?

Key question
- How do Israelis and Palestinians feel about the unequal supply of water?
- Should people always have equal access to water?

Key words
- aquifer
- desalination
- chronology
- apathy

Resources
- Activity Sheet 12
- Activity Sheet 13
- Information Sheet 8
- Information Sheet 9
- Figure 17 from CD
- Atlases

Learning objectives
- To know the location of key features in the Jordan River basin
- To understand that water supplies in the region are unequal
- To be able to argue for/against a point of view

Assessment opportunities
- Students annotate a map of the region with geographic information
- Students explain how Israelis/Palestinians might feel about water supply issues

Lesson structure

Starter
Tell students to use atlases to locate Israel and the River Jordan. Ask students to identify features that describe the basic geography of the region.

Identify and explain the status of the Palestinian Territories of Gaza and the West Bank.

> Use maps to set the geographical context for their work in the lesson. Avoid telling students which page to turn to in the atlas – insist they use their atlas index and contents pages! Prompt the students to identify basic features needed for their maps, e.g. neighbouring countries such as Syria and Jordan, capital cities and physical features such as rivers.

Main teaching phase
Hand out Activity Sheet 12 and give students time to add key geographical features to their map using the atlases.

Explain that the Jordan River is a vital source of water for Israel, Jordan and Syria. These countries already extract over 90% of the 'renewable' water the river can supply, but Palestinians extract no water from the river. Where do Palestinians get their water from? Show students the outline map of mountain aquifers (page 1 of Information Sheet 8). Whose territory do the aquifers lie under – Israeli, Syrian, Jordanian or Palestinian? Page 2 shows the outline map with a base map underlay to answer this question. Teachers notes are provided on Information Sheet 9.

> Using 'layers' of information in this way is a key principle of Geographical Information Systems (GIS). Select a method for doing so that suits your circumstances, such as drawing the location of the aquifers onto the base map using a tracing overlay or using an ICT alternative such as Google Maps to overlay the aquifers on a map of the region.

Explain that Israel gets about 495 million cubic metres per year from aquifers, Palestinians around 80 million, and that this imbalance is viewed very differently by the two groups. Arrange students in groups of four and provide each group with one set of cards and an enlarged to A3 copy of the grid from Activity Sheet 13.

Ask students to select the cards that include dates and to place these in chronological order underneath the X axis.

Next, one pair within each group should think how Israelis might react to the information and events, while the other pair imagines a Palestinian reaction. Ask each pair in turn to move the cards up the graph Y-axis to the emotion that corresponds to the Israeli or Palestinian reaction to this piece of information. They should record the position of each card with a symbol or colour (e.g. 'P' for Palestinian). How often do their reactions match each other? Why do opinions and reactions sometimes differ? In role, what reactions would they have to the fortune line cards without dates?

> Having worked as a team on the first part of the activity, students will now find they have conflicting things to say! They will need to agree a system for placing and recording the cards on the chart. Use this opportunity to encourage them to work as a team, despite their differences, and to resolve any conflicts they encounter.

Use the completed 'Fortune line' charts to elicit similarities and differences between the Israeli and Palestinian perspectives on water use. What would the two sides say about water supply in the region? What would make them feel angry, frustrated, hopeful etc.?

Plenary

In total, Israel extracts 1655 million cubic metres of water per year, the Palestinians 229 million cubic metres per year. In order to decide whether the supply of water in the region is fair or unfair, what other information would we need? How could we justify an unequal supply of water to the two groups?

Show students the 'Water' statement in Figure 17. How do they feel about this statement? Is a war over water inevitable, or can students use the more 'hopeful' cards from the previous activity to suggest ways in which the two sides could agree on the use of water?

Photo © Maurice Hopper/EAPPI/World Council of Churches

LESSON 7:
Is skiing a good use of water?

Key questions
- How can we assess sustainability?
- Why do people hold very different opinions about the use of water resources?

Key words
- sustainable
- economic
- social
- environmental
- ski resort

Resources
- Information Sheet 10
- Information Sheet 11
- Information Sheet 12
- Information Sheet 13
- Activity Sheet 14
- Figures 18-27
- Sticky notes

Learning objectives
- To know the meaning of the word 'sustainable'
- To understand that sustainability includes economic, social and environmental elements
- To be able to assess the sustainability of water use from different points of view

Assessment opportunities
- Students complete a diagram to compare points of view on a ski resort
- Students assess the sustainability of indoor skiing and explain why resorts are being built

Lesson structure

Starter
Ask the class to recall when and where they may have heard the word 'sustainability' or 'sustainable'. Record some of the responses on the board. Ask students to give some examples of sustainable behaviour and to define the term. Offer students the UN's definition of sustainable development: 'Development that meets the needs of the present without compromising the ability

of future generations to meet their own needs' and ask them to compare their own definition with this one.

Main teaching phase
Ask students to list on sticky notes their criteria for the location of a ski resort (e.g. climate, altitude, topography, access, wealth). How did they select criteria? Did they think about personal experience, physical factors or other ideas? Ask students to prioritise their sticky notes by using a diamond ranking technique (Information Sheet 10 can be displayed on the whiteboard to show this).

Students should look at their diamond ranking and justify their approach. Elicit ideas and agree on the key criteria as a class.

> 'Layered decision making' takes place when we teach students that a decision about location is not the same a making a bullet-pointed list! Some factors come before others. In this case, there can be no skiing without snow, or slopes! So why hasn't every slope with snow got a ski resort? What else is important? Infrastructure (e.g. airports and roads) needs to be in place for a resort to be built, and there has to be a wealthy set of customers. What factors have precedence?

> A 'diamond ranking' technique has one factor in the top rank, two in the rank below, three in the next rank; then two, then one factor at the bottom. This helps identify the most and least important criteria, but allows others to appear in groups of equal importance.

Show students the climate data on Information Sheet 11. Do not tell them the location at this stage! Ask them to decide in pairs whether this would be a suitable location for a ski resort.

Tell students they are going to see images from a ski resort (Figures 18-27). Show the first image and invite students to guess the location. Show the next image and allow another guess. Continue with each image and finally reveal the location: Dubai. Reveal that the climate graph seen earlier is also from Dubai and challenge students to explain how a ski resort is possible there!

Provide the background information on Ski Dubai (Information Sheet 12). Students then use Activity Sheet 14 to complete a Venn diagram comparing characters' opinions about Ski Dubai.

Allow students to choose the characters they wish to compare. Students should compare different views and consider the question 'Do our opinions of Ski Dubai depend on who we are and where we live?'.

> Venn diagrams are useful 'thinking frames' for comparing and contrasting. Identify similarities between character opinions and write these where the circles intersect. Write differences in the part of a circle outside the overlap.

Plenary

Show students the map of the location of indoor ski resorts in the UK (Information Sheet 13) and ask if they have visited any. Present students with the following data: it takes about 250,000 litres of water to create a 15cm blanket of snow covering a 50x50 metre area. Ask students:

- Why was Ski Dubai built?
- Would they ski in Dubai?
- Would they ski indoors?
- In what ways is skiing a 'good' use of water and how could it be seen as 'bad'?

> Encourage students to use information they have learned in this unit so far and to relate their answers to ideas such as our right to water, our demand for recreation, our responsibility to use resources sustainably and the need to balance economic, social and environmental considerations.

Prior to the next lesson

In preparation for the next lesson, ask students to each bring in a one-litre bottle of bottled water and to investigate how that company markets or advertises its product.

Photo © Ski Dubai

LESSON 8:

What is the cost of a bottle of water?

Key questions
- How much would you pay for a bottle of water?
- What are the impacts of drinking bottled water?

Key words
- bias
- persuade
- production
- recycle

Resources
- Bottles of water and notes on advertising brought in by students
- Cups or glasses
- One glass of chilled tap water
- Video clip 'Bottled water: is it worth it?' (http://news.bbc.co.uk/player/nol/newsid_7230000/n ewsid_7231200/7231261.stm?bw=bb&mp=wm&new s=1&ms3=6&ms_javascript=true&bbcws=2)
- Blindfold or scarf
- Information Sheet 14
- Information Sheet 15
- Information Sheet 16

Learning objectives
- To analyse the reasons people drink increasing quantities of bottled water
- To understand the impacts of drinking bottled water

Assessment opportunities
- Students identify bias and the use of persuasion in a range of images and other sources
- Students assess the social, environmental and economic costs and benefits associated with bottled water

Prior to the lesson
At the end of the previous lesson ask students to each bring in a one-litre bottle of bottled water and to investigate how that company markets or advertises its product.

Lesson structure

Starter
Ask students to discuss the selling points made in adverts for their brand of bottled water. What claims are made about the products? Why should we buy and drink them? Are the adverts aimed at a particular audience? What techniques are used to make bottled water attractive?

> Detecting bias is an essential skill in geography and is now an element of the attainment target. Encourage students to evaluate the quality of information by asking questions about its origin, e.g. who wrote the advertisement and for what purpose?

Main teaching phase
Tell students to put their litre bottles onto one desk to give a visual impression of how much water they have in total, giving them something to compare figures to during the rest of the lesson.

Ask students to estimate how many bottles of water people in the UK drink each year. What do they think has happened to bottled water consumption during their lifetime? Show page 1 of Information Sheet 14 and ask students to identify any 'rogue' figures before revealing the answer on page 2.

> 'Rogue' or false data is a way of ensuring that students engage with information, giving them a purpose to read the data. It stimulates high-order thinking and shows the need to look at information critically.

Survey the class to find out how much they paid for their litre bottle of water. Use these responses to begin a debate about an acceptable price for a bottle of water.

Tell students that Gordon Ramsay's Claridge's restaurant in London has a water list with exclusive waters from around the world. Show the flags for these countries on

Information Sheet 15 page 1 and challenge students to identify each country from its flag before revealing the answers from Information Sheet page 2. Discuss the sources of water and the prices shown.

Show the video extract 'Bottled water: is it worth it?'. Ask students to identify some of the costs associated with bottled water production and consumption. Use the newspaper article on Information Sheet 16 to identify the main costs and benefits of drinking bottled water.

> The costs are not purely financial. Ask students to consider economic, social and environmental impacts of the extraction and production, transportation, consumption and disposal of bottled water. Remind students of the use of bias. Is the information given in the clip and the news article balanced and accurate?

Ask students to decide if we are 'paying' (in the widest sense of the word) too much for our bottled water. Do the costs outweigh the benefits? Ask students to select some key images and evidence from what they have seen. How could they use their evidence to *convince* an audience that 'bottled water is worth every drop'?

Plenary
Ask for a volunteer to assist in a taste test. Blindfold them then allow them to taste a selection of different brands of of bottled water – expensive and inexpensive. Randomly introduce a cup of chilled tap water. Ask the volunteer to rank the water, or to spot the tap water!

> In the UK, drinking water must comply with 57 standards relating to its health, taste and appearance. In 1997, 99.75% of the tests met the appropriate standard. High-quality drinking water comes through the mains supply, usually the cold water tap in the kitchen, delivered uninterrupted by pipe from the supplier to prevent contamination. Typically, tap water costs around 1p per ten litres.

Homework
Explain that re-using articles uses less energy than recycling (rethink, reduce, repair, reuse, recycle!). Ask students to re-use their water bottle as creatively as they can. Discuss some possible ideas:
- used upside down as a funnel
- with the top cut off and measurements marked on as measuring jug
- filled with dried peas and sealed to make a musical instrument.

Photo © Bryan Ledgard

LESSON 9:
Will our water run out?

Key questions
- Who supplies my water?
- What do I think should happen to water use in future?

Key words
- water rate

Resources
- Activity Sheet 15
- Information Sheet 17
- Information Sheet 18
- Activity Sheet 16

Learning objectives
- To know how water is supplied and consumed in the UK
- To understand the need to balance water supply and demand
- To imagine a future with high or low levels of water use

Assessment opportunities
- Students rank statements about water use and write their own statement for the future
- Students write a letter describing a possible future for water resources

Lesson structure

Starter
In pairs, give students a set of cards from Activity Sheet 15 and ask them to arrange the statements as a 'diamond nine' – with the statement they agree with most strongly at the top, the one they disagree with most strongly at the bottom, and three layers in between. Elicit feedback from students on their opinions and reasons.

Main teaching phase
Ask students what they know about their own water supply, e.g. the name of their supplier, the source of their water and whether they have water meters or pay water rates.

> Most water companies have educational departments. You may wish to contact your local water company and arrange a visit or for an education officer to come to the school. This could form part of an investigation by students into their local water supply, or an application for Sustainable Schools status.

Show Information Sheet 17 on a whiteboard or as a hand-out. Provide each student with the water company mission statements from Information Sheet 18 and ask them to identify from these statements the main issues facing water companies. Is meeting demand (increasing supply) more important than reducing it (encouraging water conservation)?

Ask students to imagine what could happen to water demand and supply in their local area by 2050. What might the local water company mission statement say then?

Ask students to draft a letter set in the year 2050, describing one of two possible futures:

1. One where demand for water has risen unchecked. How would water be used? Where would it come from? What might the impacts be – on costs and on the environment?

2. One where water is carefully conserved. How might life be different? How would water conservation be achieved? How much would it cost and would it be affordable for all?

Plenary

Allow some time for individual reflection on any impact this unit has had. What are the key ideas learned? Have students changed their attitudes towards the use of water? What future would they like to see for the way in which water resources are used locally and globally? Give each student a copy of Activity Sheet 16 and ask them to write their own personal message about water on the 'message in the bottle', then cut out the bottle and add them to a class wall display. Allow time for students to read one another's messages and discuss as appropriate.

Young people benefit from developing a futures perspective on their lives and on the world. The purpose of such a dimension is to help teachers and students:
- develop a future-orientated perspective on their lives and events in the world
- exercise critical thinking skills and use creative imagination more effectively
- identify alternative futures which are more just and sustainable
- engage in responsible citizenship in the local, national and global community, on behalf of both present and future generations (Hicks, 2001).

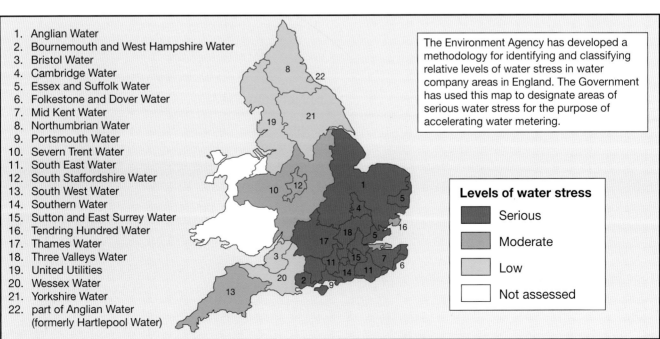

1. Anglian Water
2. Bournemouth and West Hampshire Water
3. Bristol Water
4. Cambridge Water
5. Essex and Suffolk Water
6. Folkestone and Dover Water
7. Mid Kent Water
8. Northumbrian Water
9. Portsmouth Water
10. Severn Trent Water
11. South East Water
12. South Staffordshire Water
13. South West Water
14. Southern Water
15. Sutton and East Surrey Water
16. Tendring Hundred Water
17. Thames Water
18. Three Valleys Water
19. United Utilities
20. Wessex Water
21. Yorkshire Water
22. part of Anglian Water (formerly Hartlepool Water)

The Environment Agency has developed a methodology for identifying and classifying relative levels of water stress in water company areas in England. The Government has used this map to designate areas of serious water stress for the purpose of accelerating water metering.

Levels of water stress
- Serious
- Moderate
- Low
- Not assessed

4: GLOSSARY

apathy – lack of interest or enthusiasm

aquifer – layer of porous rock storing useful amounts of water

bias – prejudice towards a particular viewpoint

chronology – arrangement of events/dates in order of occurrence

citizen – a member of a state or inhabitant of a place, with rights

condensation – turning from gas to liquid (e.g. water vapour to liquid water droplets)

conflict – a struggle between two or more opposing forces, ideas or interests

conserve – protect or save, e.g. the earth and its resources

consumption – the total amount of any resource/product used

co-operation – acting or working together with others to achieve a common goal such as the resolution of conflict

dam – a structure across a river retaining water in a lake or reservoir, used to store water, regulate floods and/or generate power

desalination – the process of removing salt from sea water (for human use)

drainage basin – area of land drained by a river and its tributaries

economic – associated with money or work

environmental – associated with the physical and natural surroundings

evaporation – turning from liquid into gas (e.g. water to water vapour)

glacier – a moving mass or 'river' of ice formed by accumulated and compacted snow

groundwater – water stored underground in permeable rock

human right – a basic right and freedom to which all people are entitled, including the right to life and liberty, freedom of thought and expression, and equality before the law

hydrosphere – the parts of the earth containing water, e.g. seas, rivers, glaciers, clouds

ice sheet – slowly moving mass of ice covering a wide surface area

need – a necessary requirement to sustain life, without which we cannot live, e.g. water

peace accord – a formal agreement between states to end a conflict

persuade – convince others to think in a certain way

production – the process of making/manufacturing an object

quality of life – the wellbeing of people measured by many economic and social factors

recycle – treat waste products for later re-use

resource – something that is useful for people

responsibility – having control, being accountable or being the cause of something

scarcity – a lack or inadequate amount, e.g. of water

ski resort – an area developed for skiing including accommodation and facilities

social – associated with people and their quality of life

soil moisture – the amount of water held in the soil

spring – water emerging naturally to the surface from the ground

surplus – the amount left over after what is needed has been used

sustainable – meeting the needs of the present without compromising the ability of future generations to meet their own needs

United Nations (UN) – an international organisation composed of most countries of the world, founded in 1945 to promote peace, security and economic development

United Nations Environment Programme (UNEP) – the UN body that encourages sustainable development in developing countries

virtual water – 'embedded water' used to produce products from beer to cars

want – something that is desired but is not essential to life

water balance – the amount of water reaching an area via rain and rivers less the amount lost through evaporation and river outflow

water cycle – the circulation of water between the atmosphere and the Earth's surface

water dependency – the percentage of renewable water originating outside a country

water footprint – the total amount of water used and consumed by an individual both directly and indirectly

water rate – a fixed or 'standing' charge for water that is not dependent on its use

water resource – water available for human use

water supply – the provision and storage of water for human use

watershed – dividing line between the drainage basins of rivers, often a ridge of higher land

West Bank – a region west of the River Jordan and north-west of the Dead Sea. Of its inhabitants 97% are Palestinian Arabs

5: LINKS FOR FURTHER IDEAS AND RESOURCES

References and further reading

Caldecott, J. (2007) *Water: Life in every drop*. London: Virgin Books.

Clarke, R. and King, J. (2004) *The Atlas of Water*. London: Earthscan.

Falkenmark, M., Berntell, A., Jägerskog, A., Lundqvist, J., Matz, M. and Tropp, H. (2007) *On the Verge of a New Water Scarcity: A call for good governance and human ingenuity*. Stockholm: SIWI.

Hicks, D. (2001) *Citizenship for the Future. A practical classroom guide*. Guildford: WWF-UK

International Commission on Large Dams (2007) *Dams and the World's Water*. Paris: ICOLD.

Leat, D. and Nichols, A. (1999) *Theory into Practice: Mysteries make you think*. Sheffield: Geographical Association.

McPartland, M. (2001) *Theory into Practice: Moral dilemmas*. Sheffield: Geographical Association.

McPartland, M. (2006) 'Strategies for approaching values education' in Balderstone, D. (ed) *Secondary Geography Handbook*. Sheffield: Geographical Association, pp. 170-79.

Morgan, A. (2006) 'Teaching geography for a sustainable future' in Balderstone, D. (ed) *Secondary Geography Handbook*. Sheffield: Geographical Association, pp. 276-95.

Nichols, A. (2006) 'Thinking skills and the role of debriefing' in Balderstone, D. (ed) *Secondary Geography Handbook*. Sheffield: Geographical Association, pp. 180-97.

Pearce, F. (2006) *When the Rivers Run Dry*. London. Eden Project books.

Raines Ward, D. (2002) *Water Wars: Drought, flood, folly and the politics of thirst*. New York, NY: Riverhead books.

Roberts, M. (2003) *Learning Through Enquiry: Making sense of geography in the key stage 3 classroom*. Sheffield: Geographical Association.

Scoones, S. and Willson, A. (2002) *Whose Right to Water?* London: Worldaware.

Watkins, K. (2006) *Human Development Report: Beyond Scaricty: Poverty and the global water crisis*. New York, NY: Palgrave McMillan.

Wellsted, E. (2006) 'Understanding distant places' in Balderstone, D. (ed) *Secondary Geography Handbook*. Sheffield: Geographical Association.

Websites

Environment Agency
News, information, research and data about environment of England and Wales
www.environment-agency.gov.uk

Envirowise
Advice for businesses on how to reduce costs by reducing water consumption
www.envirowise.gov.uk/towaternet

SAPERE
A guide to Philosophy for Children (P4C) in the UK, created by an educational charity interested in the role of philosophical enquiry in education as a model of rigorous thinking and as a celebration of wonder and open-mindedness
http://sapere.org.uk/

The Right to Water
Established by WaterAid/Rights and Humanity, in co-operation with Freshwater Action Network, focuses on water as a human right and on access to a clean and reliable water supply for everyone.
www.righttowater.org.uk

The World's Water
A project of the Pacific Institute for Studies in Development, Environment and Security, based in California, provides water information, data and resources to individuals, organisations, and institutions working on global freshwater problems
http://www.worldwater.org/

WaterAid
A UK development charity that works to provide
deprived communities in Asia and Africa with a safe
water supply, sanitation and associated hygiene
practices
www.wateraid.org.uk

Water for Life
The website for the UN International Decade for Action
Water for Life (2005-2015), covering water scarcity, access
to sanitation, health, water resource management and
trans-boundary water issues
www.un.org/waterforlifedecade

Water in Schools
A range of resources created by Thames Water for
teachers and students to support teaching and learning
about water
www.waterinschools.com

Water UK
Working on behalf of the water industry towards a
sustainable future
www.water.org.uk

Waterwise
An independent, not-for-profit organisation that receives
funding from the UK water industry, sponsorship and
consultancy, focused on reducing water consumption
and increasing water efficiency in the UK
www.waterwise.org.uk

6: ASSESSMENT FRAMEWORK:
Do we have equal rights to resources?

Level 7-8

Students can:
- identify and analyse patterns of water use at a range of scales from personal to global

Level 5-6

Students can:
- explain the importance of water in their own lives and recognise similarities and differences in relation to other people's consumption of water

Level 3-4

Students can:

- describe human use of water resources (including their own)

- recognise the link between water, human rights and quality of life

- recognise and describe patterns of water supply and demand

- offer reasons for their views on water resource conflicts and their solutions

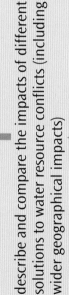

- describe and compare the impacts of different solutions to water resource conflicts (including wider geographical impacts)

- evaluate sources of evidence critically and recognise bias and opinion in data

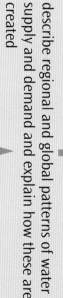

- describe regional and global patterns of water supply and demand and explain how these are created

- understand sustainable and other approaches to water resource management and the unintended effects of human actions

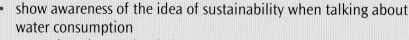

- show awareness of the idea of sustainability when talking about water consumption
- appreciate the need to balance water rights with responsibilities

- make substantiated arguments about the solutions to water conflicts and suggest how water should be managed in the future

PoS coverage in the *Toolkit* series

		Into Africa	Rise and Rise of China	British or European?	Look at it this way	Water works	Thorny issue	Faster, higher, stronger	Changing my world	Moving stories	Future floods
KEY CONCEPTS	Place	✓	✓	✓	★	★	★	✓	★	★	✓
	Space	★	★	★	✓	✓	★	✓	★	✓	
	Scale	★	★	★	★	★	★	★	✓	★	★
	Interdependence	✓	✓	★		★	✓		★	★	
	Physical human process		★		✓	★	✓	✓	✓	✓	★
	Environmental interaction	★	✓		✓	✓		★	✓		✓
	Diversity	✓		✓	★		★	★		✓	
KEY PROCESSES	Enquiry	✓	✓	✓	✓	✓	✓	✓	✓	✓	✓
	Fieldwork				★						✓
	Graphicacy	★	★	✓	★	★		★	★	✓	✓
	Communication	★		★	★	★	★	✓	★	★	
RANGE AND CONTENT	Variety of scale	★		★		★		★	✓	★	
	Location	★	★	✓	★		★	★			★
	Aspects of UK			✓	★			✓	★	✓	✓
	Parts of the world	✓	✓	✓		✓	✓		★		
	Physical geography				✓	★	★				★
	Human geography	✓	★	★				★		✓	
	People-environment	★	✓		✓	✓	★	★	✓		✓
CURRICULUM OPPORTUNITIES	Personal experience	★		✓	★	★	★		★	★	
	Contemporary context	✓	✓	★	★	★	★	✓	★	★	★
	Enquiry approaches	★	★	★	★	★	✓	★	★	★	
	Maps & GIS	★	★	★	★	★	★	★	★	★	✓
	Fieldwork				★						✓
	Responsible action	★		✓		★	✓	★	✓		
	Issues in the news	★	✓	★	★	✓		✓	✓	✓	✓
	Use of ICT				✓		★	★	★		★
	Curriculum links			★	★	★	★	★		✓	

KEY: ✓ major focus/fully developed ★ additional aspect